D0786607

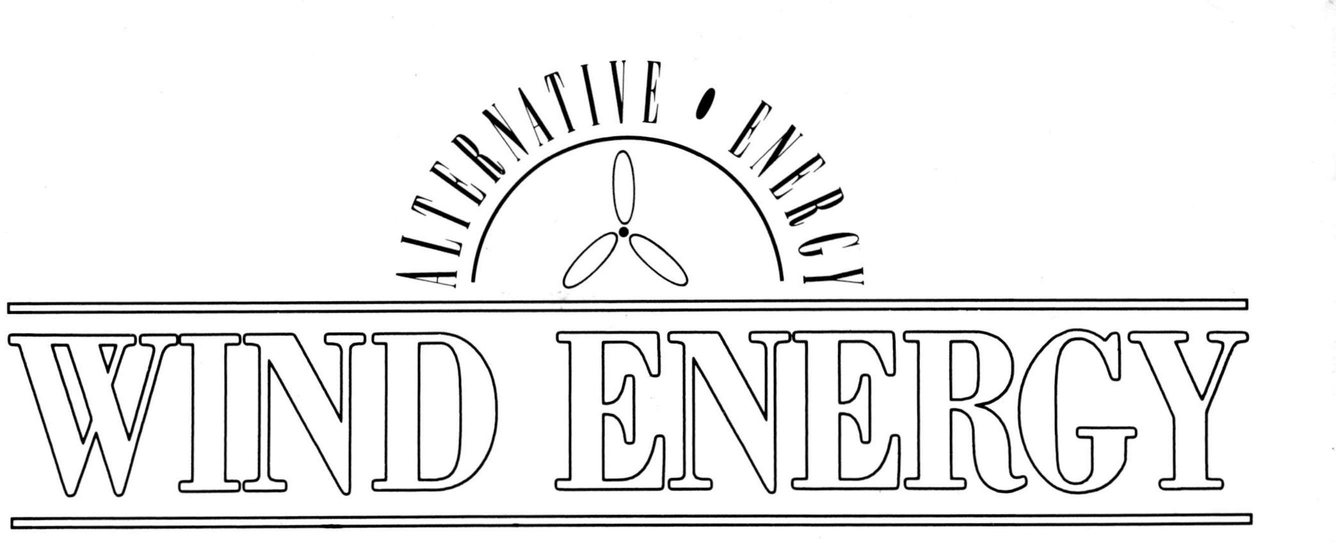


GRAHAM RICKARD

Gareth Stevens Children's Books
MILWAUKEE

**Titles in the Alternative Energy series:**

*Bioenergy*
*Geothermal Energy*
*Solar Energy*
*Water Energy*
*Wind Energy*



**For a free color catalog describing Gareth Stevens' list of high-quality children's books, call 1-800-341-3569 (USA) or 1-800-461-9120 (Canada).**

**Library of Congress Cataloging-in-Publication Data**

Rickard, Graham.
Wind energy / Graham Rickard.
p. cm. — (Alternative energy)
"First published in the United Kingdom, copyright 1990, by Wayland (Publishers) Limited"—T.p. verso.
Includes index.
Summary: Discusses how today's modern wind turbines have developed from the windmills of the past and how technology relating to wind power continues to develop.
ISBN 0-8368-0711-1
1. Wind power—Juvenile literature. [1. Wind power.] I. Title. II. Series: Alternative energy (Milwaukee, Wis.)
TJ820.R53 1991
621.4'5—dc20 91-9263

North American edition first published in 1991 by

**Gareth Stevens Children's Books**
1555 North RiverCenter Drive, Suite 201
Milwaukee, Wisconsin 53212, USA



**Picture acknowledgements**

Artwork by Nick Hawken

The publishers would like to thank the following for supplying photographs: Biofotos (Heather Angel), 23; David Bowden, 5 (left), 17 (right); British Aerospace/Wind Energy Group, 16, 19 (both), 20; Danish Wind Energy Association, 24; Energy Technology Support Unit, 26 (right); Eye Ubiquitous, 8, 13, 26 (left); Greenpeace, 4, 5 (right); Japan Ship Center, 25; Christine Osborne, 10, 12; Topham, 9, 11; U.S. Department of Energy, cover, 17 (left), 18, 21.

Editors (U.K.): Paul Mason and William Wharfe
Editors (U.S.): Eileen Foran and Kelli Peduzzi
Designers: Charles Harford and David Armitage
Consultant: Jonathan Scurlock, Ph.D.

Printed in Italy

1 2 3 4 5 6 7 8 9 95 94 93 92 91

# Contents



Words that appear in the glossary are printed in **boldface** type the first time they appear in the text.

# WHY ALTERNATIVE ENERGY?

Energy is the ability to do work. All animals and plants need energy in order to live. All machines need energy to make them work. As the world's population increases and people use more machines, more energy is needed to power them. The world's demand for energy has increased by more than ten times since the beginning of the twentieth century.

Most of this energy is produced by burning one of three **fossil fuels** — oil, natural gas, and coal. But there is only so much fossil fuel in the world. Fuel supplies that took millions of years to build up are being burned at the rate of over half a million tons an hour. At this rate, all of the world's oil and gas will be gone by the year 2040.

Even worse, fossil fuels cause serious damage to the Earth's environment. When they burn, fossil fuels produce poisonous gases that turn into **acid rain**. Acid rain pollutes vast areas of the world, killing trees, fish, and wildlife. Some of these gases also contribute to what we call the **greenhouse effect**, which is gradually warming up the Earth's **atmosphere**.

*Many industries pollute the environment. They burn fossil fuels and pour poisonous gases from their chimneys into the atmosphere.*

Because of all these problems, people all over the world are looking for alternative sources of energy. Some people see the use of **nuclear energy** as the best alternative to fossil fuels. But nuclear energy depends on supplies of uranium, which is even rarer than the fossil fuels we now use. Also, the pollution caused by nuclear energy is far more dangerous than anything produced by fossil fuels. So scientists and environmentalists are working together to come up with safe, clean, and renewable sources of energy.

*The banner on this chimney was part of a campaign against acid rain. Four chimneys each carried one letter of the word "STOP."*

Many natural sources of energy are all around us. The power of the Sun, wind, and the movement of tides and rivers all provide clean, renewable energy, if we can only come up with ways to use it. This book looks at the different ways of using wind energy. Wind is a renewable energy source. Unlike coal or oil, it will never run out, no matter how much we use it. Some countries already use wind energy to generate electricity. As scientists find even better ways of using wind energy, more and more people will use it. In the future, the wind will be one of the world's main sources of energy, because it is safe, cheap, and clean.

*Sports such as windsurfing depend on the wind to provide power.*

# THE POWER OF THE WIND

The energy in the wind comes from the Sun. The Sun heats the air around the Earth. In some areas, the air is heated more than in others. For instance, the air at the equator is much warmer than the air at the North Pole.

Warm air is lighter than cold air. So, as the air is warmed up by the Sun, it rises. Cooler air moves in to replace the warm air that has risen. This cold air is then warmed up, and it too rises, to be replaced by yet more cold

*Wind is formed when warm air rises and cooler air moves in to replace it. This happens all over the Earth.*

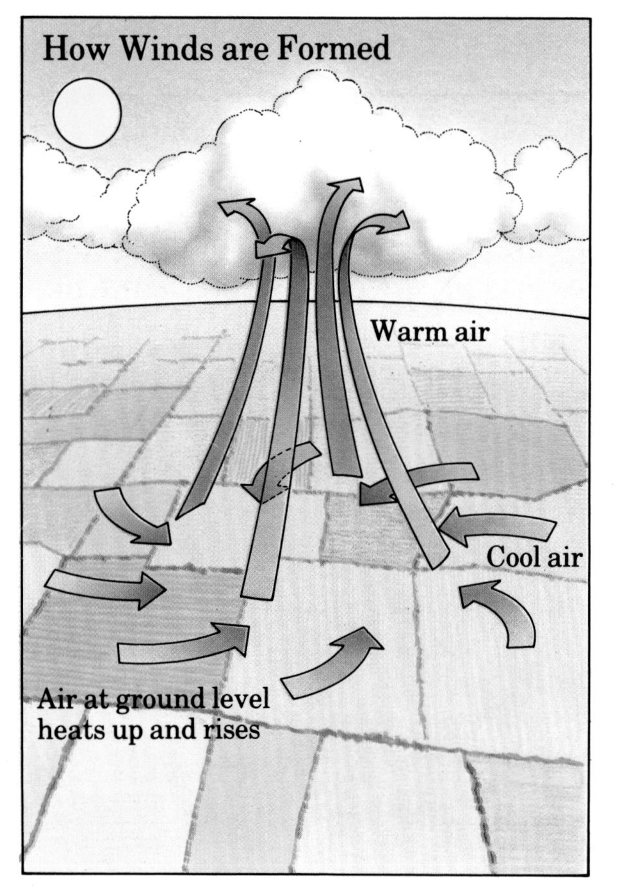


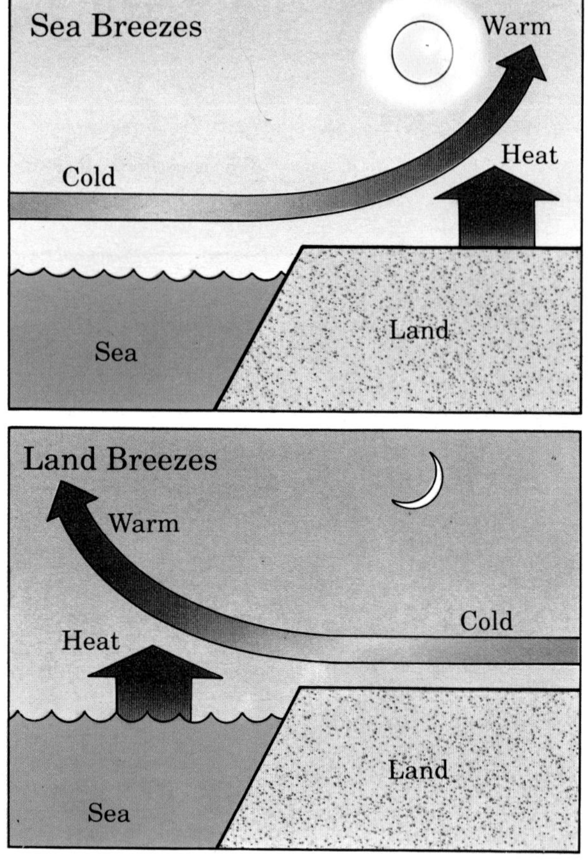


*Sea and land breezes make the coasts much more windy than other areas. This makes coasts good places for using wind energy.*

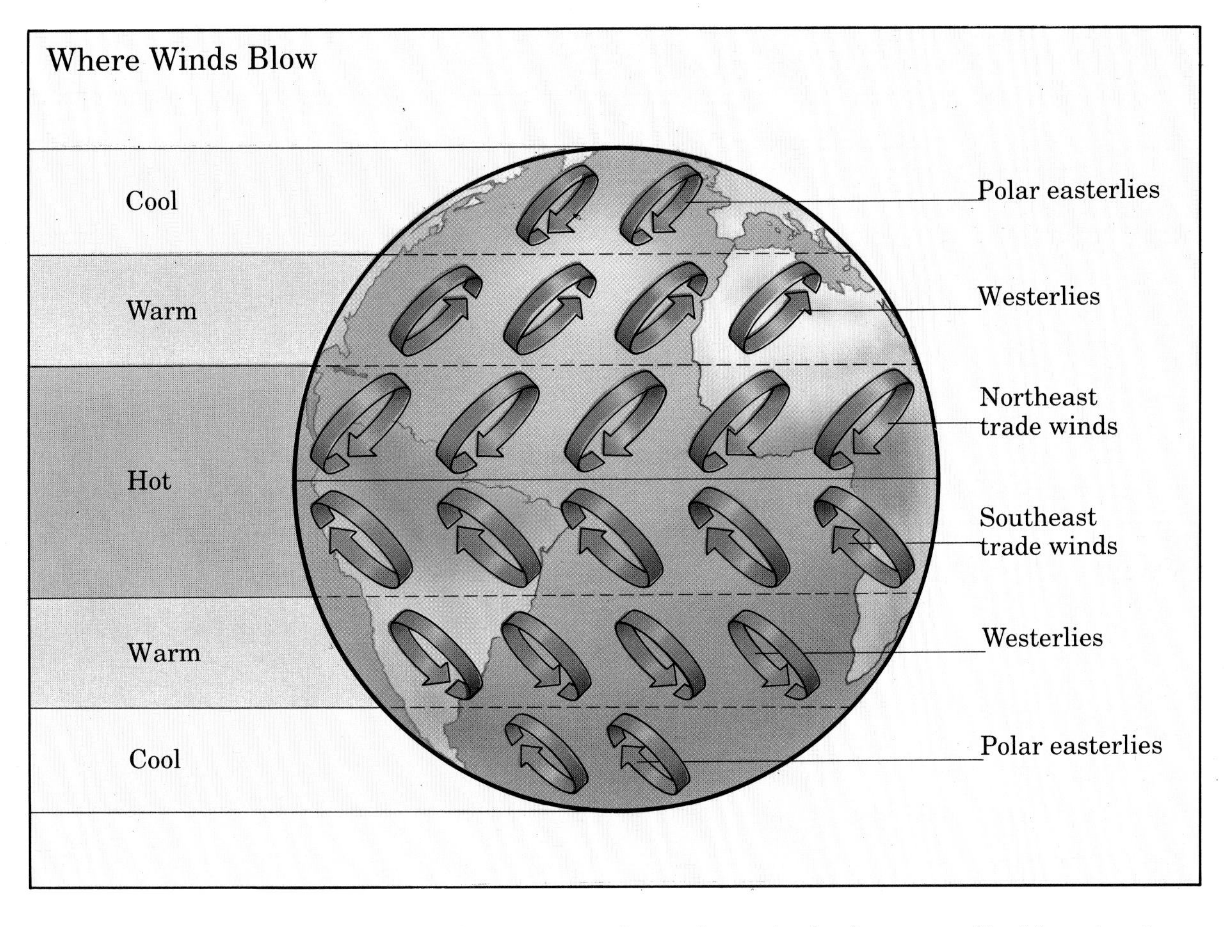


*Over the whole world, there is a pattern of regular winds that usually blow in the same direction. Winds are caused by the Sun's heat.*

air. These movements of the air form the major winds that blow over the surface of the Earth all the time.

Winds are also caused by the way that the Sun heats up the land more quickly than the sea. During the day, coastal breezes are caused by cold air flowing from the sea to replace the air that has been warmed over the land. At night, the process is reversed. The land cools down faster than the sea, and so the wind blows in the opposite direction.

Winds can vary in strength, from the lightest breeze to the fierce, terrifying power of a hurricane or tornado. The faster the air moves from one place to another, the stronger the wind. So a high wind speed means a strong wind. To measure the wind speed, scientists use an **anemometer**, which has two or three cups to catch the wind. These cups revolve on an upright spindle. The higher the wind speed, the faster the cups turn. A hurricane is a wind that blows at least 74 miles (119 km) per hour!

*An anemometer is used to measure wind speed. The harder the wind blows, the faster the cups spin. They turn a spindle. The spindle is attached to a speedometer that shows the wind speed.*

The moving air in any wind contains large amounts of **kinetic energy**. This kinetic energy can be used to do things that would otherwise require a lot of work. A windmill, for example, is a machine that uses wind energy to grind grain. Wind energy can also be used to produce electricity, a very concentrated and convenient form of energy.

Wind energy is free, uses no fuel, and produces no pollution. The strongest winds usually occur in winter, when we most need power to provide light and heat. If wind could be used efficiently, it could provide people around the world with ten times more energy than they now use.

Wind energy does have some disadvantages, however. Winds vary in direction and speed from place to place, from season to season, and even at different times of day. Therefore, we cannot depend on wind energy in a single place as the only source of power.

In spite of these problems, wind energy is one of the most promising alternatives to fossil and nuclear fuels. Someday, these fuels will be replaced by several different forms of renewable energy. In the future, wind power may supply the world with as much as 20 percent of its energy.

*After a severe hurricane in Texas, this man points to a heap of wreckage, which is all that is left of his home.*

# WIND POWER IN THE PAST

In earliest times, if people needed to get a job done, they did it themselves or used an animal, such as an ox, to help them. Then, around 5,000 years ago, they discovered how to harness the wind to do work. They learned how to move a small boat by making sails from tree bark or animal skins. The wind filled the sails and pushed the boat along.

Sailing was much easier than rowing, and the idea soon spread. Using sailing ships, people could travel long distances, and sea trade began. By the end of the eighteenth century, people had

*These Arab sailing boats are from Abu Dhabi. They look much like the boats that were used centuries ago in this region.*

*An old sailing ship, called a "clipper," in full sail.*

built large fleets of mighty sailing ships, called **clippers**.

Driven by the **trade winds**, clippers carried tea, coffee, sugar, spices, cotton, and other goods around the world. This great age of sailing ships ended with the invention of the steam engine, and later the **diesel** engine. Ships with engines were faster than sailboats and did not have to rely on the ever-changing wind.

Over 2,000 years ago, people discovered that the wind was useful on land as well as at sea. The first windmills that we know of were used for grinding grain in **Persia** around 200 B.C. These windmills had sails made from bundles of reeds, which caught the wind and turned an upright pole connected to a **grindstone**. Later, a new kind of windmill was invented. It had sails that turned like an airplane propeller. This kind of windmill had cloth sails attached to a round frame. The sails were fixed at the top of a stone tower.

*Windmills like this one first appeared in Persia.*

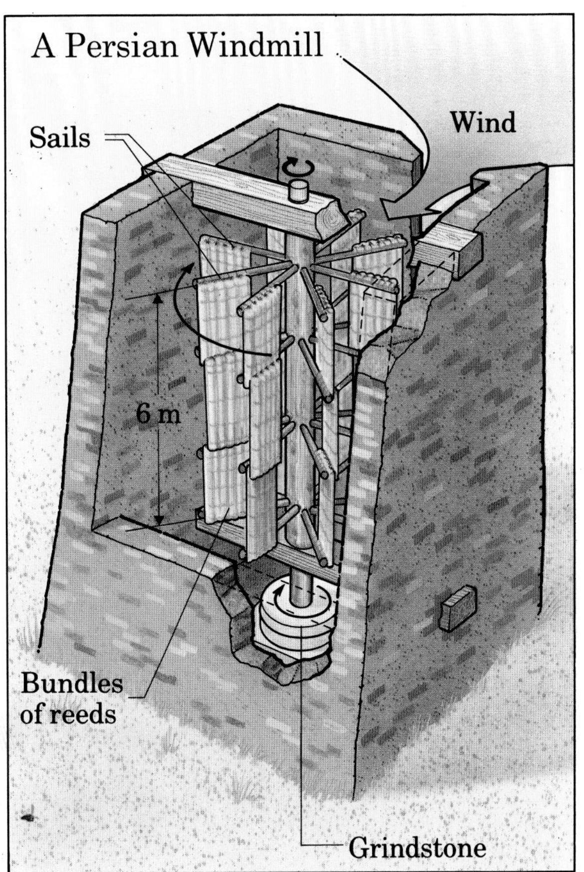

*A traditional Greek windmill on the island of Kos. The sails are very much like those used on boats. You can see that the wind has caught the sails.*

These early windmills had one problem. If the wind was blowing from the wrong direction, the windmill would not work. This problem was solved with the invention of the post mill. On the outside of post mills was a long wooden beam, which the **miller** used to turn the entire windmill into the wind when it changed direction. This must have been very hard work, but it was a great improvement over early mills.

Turning the sails into the wind was made easier with the invention of the tower mill around A.D. 1400. Tower mills were first used in Holland and soon spread to the rest of Europe. This type of mill had a fixed tower, but the sails were attached to a movable top section that could be turned into the wind. This was much simpler than turning the whole windmill. In 1745, British inventor Edmund

Lee made the miller's work even easier with his invention of the fantail. This was a miniature windmill, mounted sideways, opposite the main sails (see diagram below). If the wind changed direction, the fantail turned the top section until the main sails faced the wind again.

Post mills and tower mills were built in many countries. For several centuries, they were the main source of energy for a wide variety of jobs, such as sawing wood, pumping water, and making paper. Then, like sailing ships, windmills fell into disuse after the introduction of steam and diesel engines. These engines were more powerful, efficient, and reliable than windmills. Holland, for example, is famous for its windmills and had 12,000 of them by 1800. But by 1960, Holland had fewer than 1,000 windmills in working condition.

*The fantail keeps the main sails turned into the wind.*

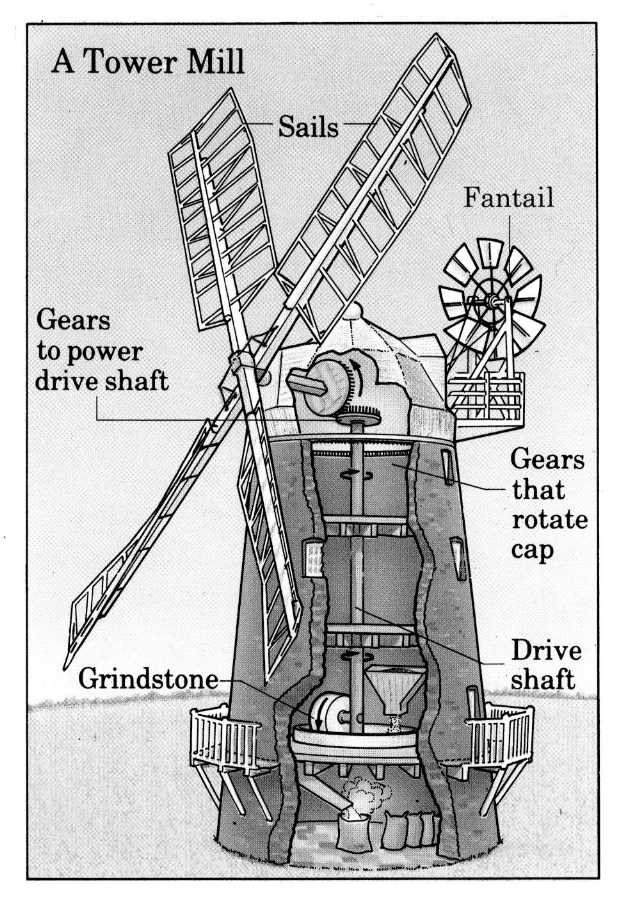

*This post mill in Holland is used for sawing wood.*

# MODERN WIND POWER

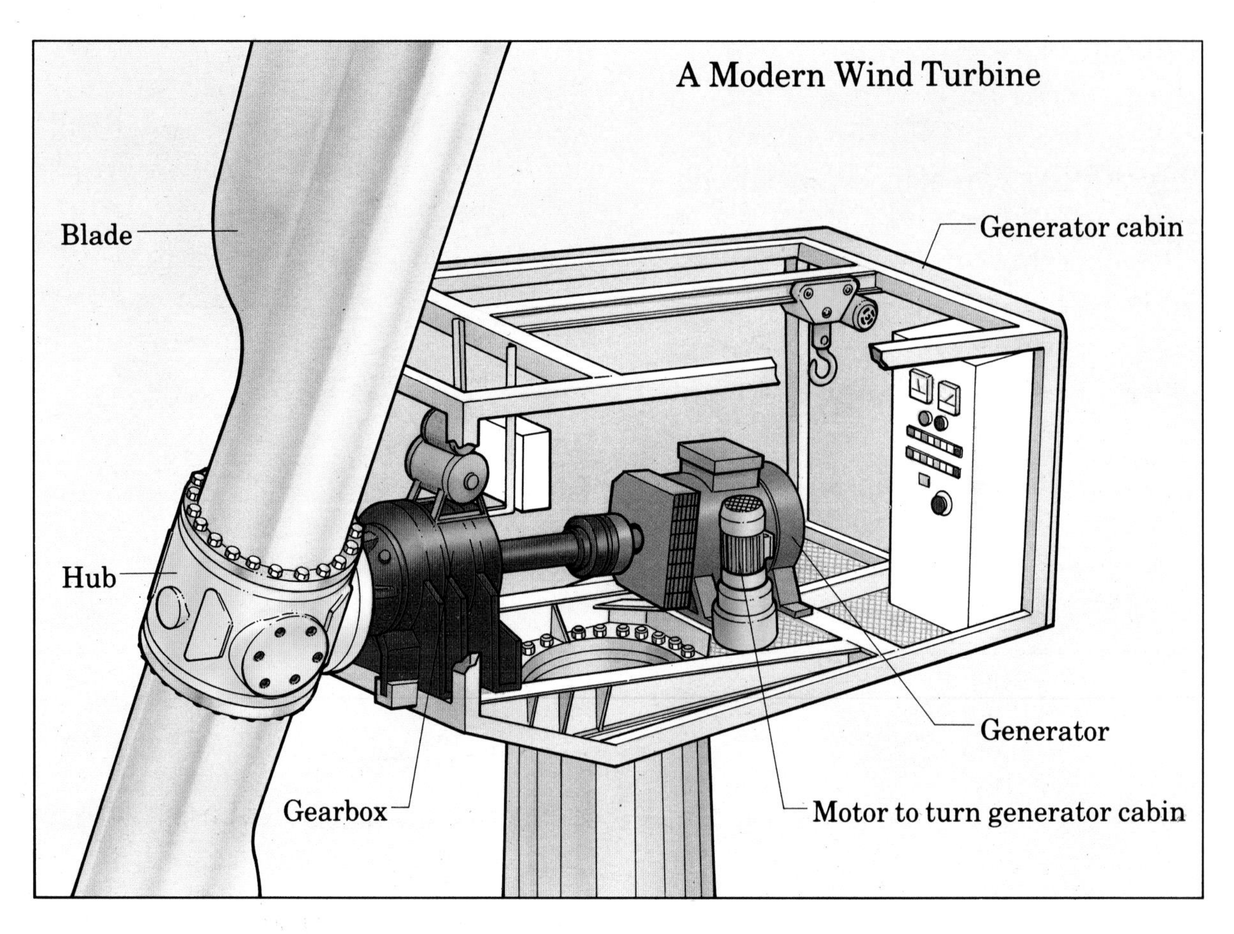


*The inside of a modern wind **turbine**, which produces electricity by capturing the wind's energy to drive a **generator**.*

People all over the world are showing a new interest in wind energy. Instead of grinding grain, however, most modern wind machines generate electricity. Electricity can be used to provide heat, light, and power. Wind machines that produce electricity are called wind turbines.

Wind turbines were first built in Denmark in the 1890s. They were used to provide electricity to places that did not receive electricity from power stations. At one time, these wind turbines were able to provide 25 percent of all the electricity used by Danish industry.

During the 1920s and 1930s, remote farms in Australia and the United States had their own wind-powered generators. In the United States, wind turbines were also used to provide electricity for interstate radio communications.

The world's first large wind-powered generator was built in the United States in 1947. However, it was almost 30 years before people began to treat wind energy as a real alternative to fossil fuels.

There are two main types of wind turbines — **horizontal-axis turbines** and **vertical-axis turbines**. Horizontal-axis machines have a turbine mounted on top of a tall tower. Traditional windmills have four or more sails, but wind turbines usually have two or three blades. These have curved surfaces, like the wings of an airplane. As the wind pushes against the blades, it turns the turbine. The turbine is connected to a generator. As the turbine turns the generator, it produces electricity.

*Of all the different types of wind turbines, the most popular are the horizontal-axis styles.*

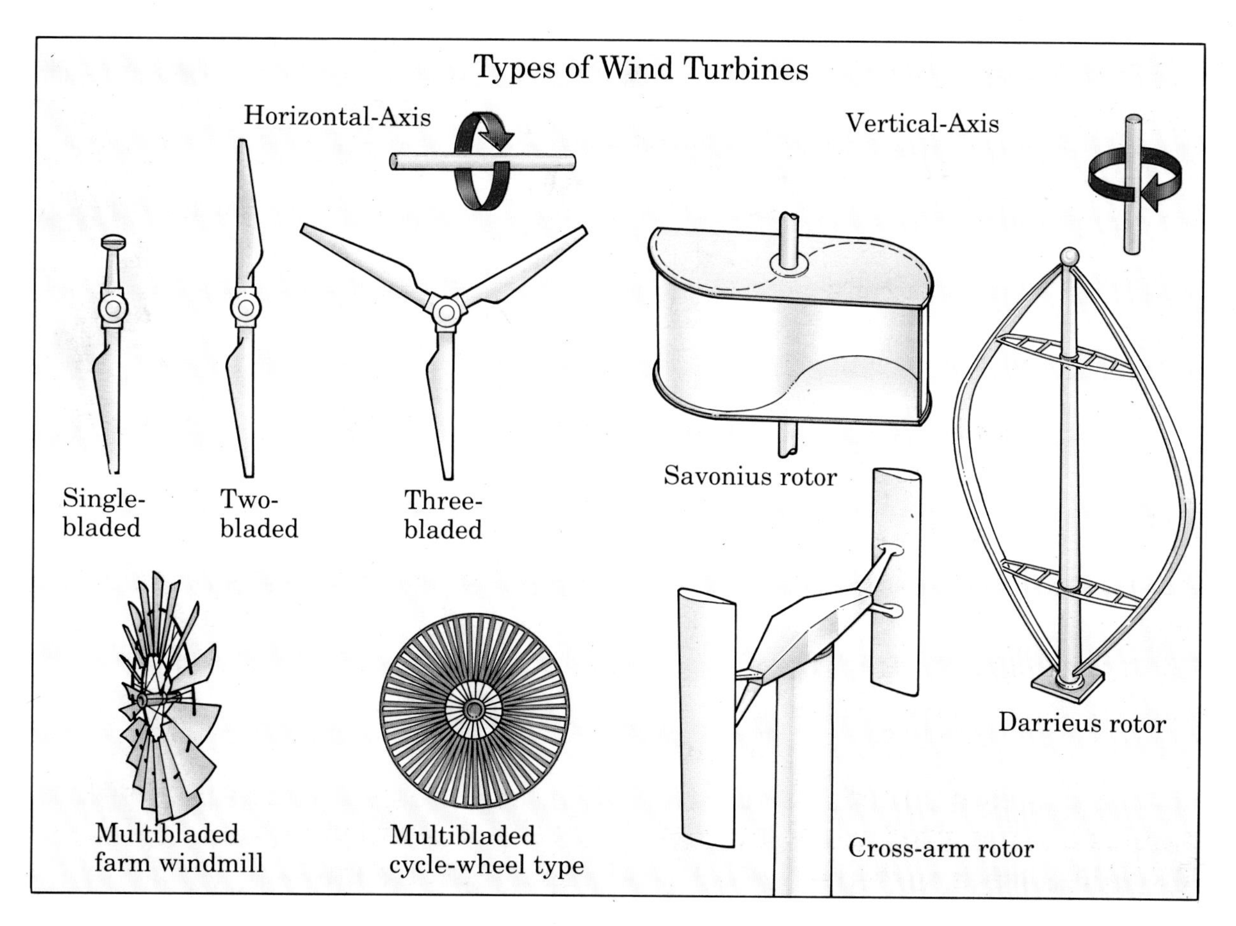

*The enormous blades of the British LS-1 wind turbine being tested. The blades were later fitted on the turbine tower on Orkney Island (see page 19).*

The blades of a turbine must be light enough to turn easily, but strong enough to stand up to fierce gales. Several materials have been tried for the blades, including wood, steel, **fiberglass**, and **carbon fiber**. Some turbines are controlled by a computer, which automatically keeps the turbine turned into the wind. Turbines can also adjust to changing wind speeds by changing the angle of their blades as they face the wind. If the turbine spins too fast, however, it can break. So, in very high winds, automatic brakes slow the turbine.

Vertical-axis wind turbines do not look like traditional windmills at all. These turbines have several different designs. The Savonius turbine, for example, was invented by a Finnish engineer. It has two halves of a **cylinder** mounted on an upright shaft. The Darrieus turbine has thin metal blades that are fixed to an upright shaft at both ends. (See page 15.)

*A large Darrieus turbine being tested in California.*

*Wind pumps are often used to pump water from rivers and lakes.*

Other turbines, such as the Giromill, change their shape to adjust to the wind speed. Vertical-axis machines have some advantages over other types; they don't need a tall tower to support them, and they work no matter which way the wind blows. But they are not as efficient as horizontal-axis machines.

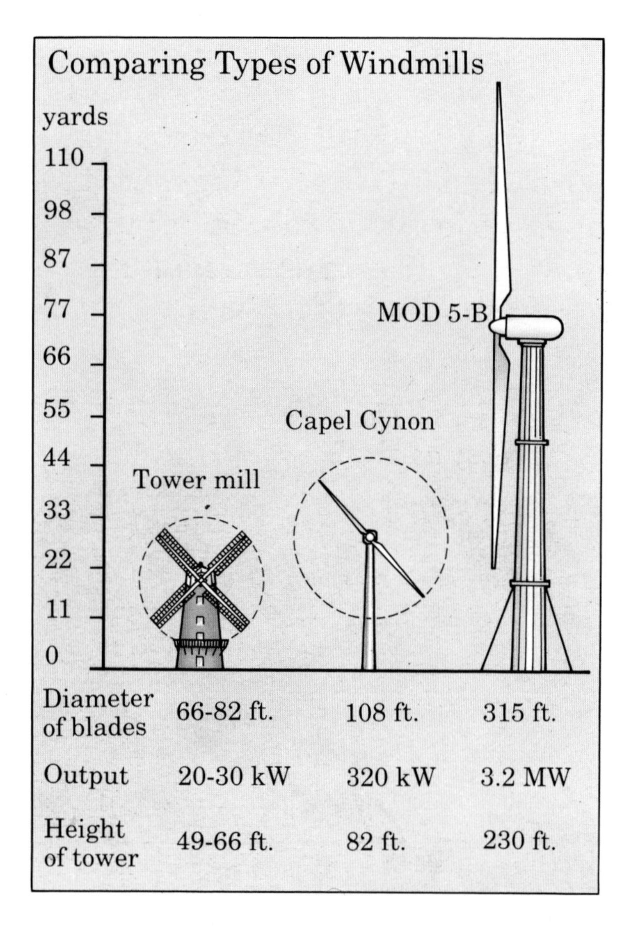


| | Tower mill | Capel Cynon | MOD 5-B |
|---|---|---|---|
| Diameter of blades | 66-82 ft. | 108 ft. | 315 ft. |
| Output | 20-30 kW | 320 kW | 3.2 MW |
| Height of tower | 49-66 ft. | 82 ft. | 230 ft. |

*Modern wind machines are more powerful than old windmills.*

Modern turbines are much more powerful than traditional windmills. The amount of electricity a turbine can produce depends on the strength of the wind and the size of the blades. If you double the length of the blades, you increase by four times the amount of electricity that they can produce.

Small and medium-sized wind turbines are already supplying electricity to many islands and other remote areas throughout the world. For example, a 200-**kilowatt (kW)** turbine on Cuttyhunk Island, off the coast of Massachusetts, provides electricity for 200 people. This wind turbine has cut the amount of fossil fuel used by the people of Cuttyhunk in half.

Some countries, including the U.S., Britain, Denmark, and Canada, have built wind turbines capable of producing more than one **megawatt (MW)** of electric

*One of the first large wind turbines in action in California.*

power (enough to boil 500 tea-kettles of water at the same time).

Two of the biggest wind turbines in the world are the British LS-1 on the island of Orkney, and the American MOD 5-B in Hawaii.

The blades of the LS-1 are almost 66 yards (60 m) in diameter, and it produces 3 MW of electricity. The MOD 5-B is even bigger. Its blades are 105 yards (96 m) in diameter. It produces enough electricity for 2,000 people.

*The MS-1 (left) was one of the first large wind turbines. The newer and bigger LS-1 (right) produces twelve times as much electricity as the MS-1.*

# FARMING THE WIND

Even very large wind turbines can only produce enough electricity to supply a small town. For large-scale production of electricity, many hundreds of turbines are grouped together to form wind farms.

Experimental wind farms have been set up in many countries, but the state of California has more than anyplace else. The California wind farms are located in a range of **mountain passes**. The wind farms include 300 different types and sizes of wind turbines. California wind farms have been used as test sites for new turbine designs from all over the world.

*Experimental wind turbine designs from all over the world are tested at the Altamont Pass wind farm in California, where the wind conditions are ideal.*

*Darrieus turbines produce electricity on a wind farm in the Tehachapi region of California. These turbines work in any wind direction.*

With over 16,000 turbines, these wind farms account for 85 percent of all wind turbines. It would take over 79 million gallons (300 million L) of oil in a power station to produce as much electricity as California's wind farms generate in one year.

Many wind farms are part of a power **network**. On windy days, the wind farm owners are paid for the electricity they produce. The electricity is fed into the network. In return, they take power from the network on calm days when the wind can't turn their turbines. Farmers in Denmark and the U.S. have also made money from wind farms. Since wind power is a clean form of energy, their land can still be used for ordinary farming.

## FARMING THE WIND

To make the most of a wind farm, it must be placed in a suitable area. The best places are coasts, high plains, or mountain passes, where the wind blows most of the time. Before building a wind farm, scientists measure the wind speed and direction for several years. Then they are able to decide on the size and number of wind turbines that would make full use of the wind.

At least a dozen countries are already experimenting with wind farms. India, China, and many European countries have ambitious plans for the future. Wind farms are relatively expensive to build, but they cost very little to run, so they produce electricity as cheaply as conventional power stations. Wind farms do not use fuel, and they do not produce pollution.

*This is what the wind farm at Capel Cynon in Wales will look like when it is completed. This wind farm is the first in Britain. It will produce enough electricity for about 5,000 people.*

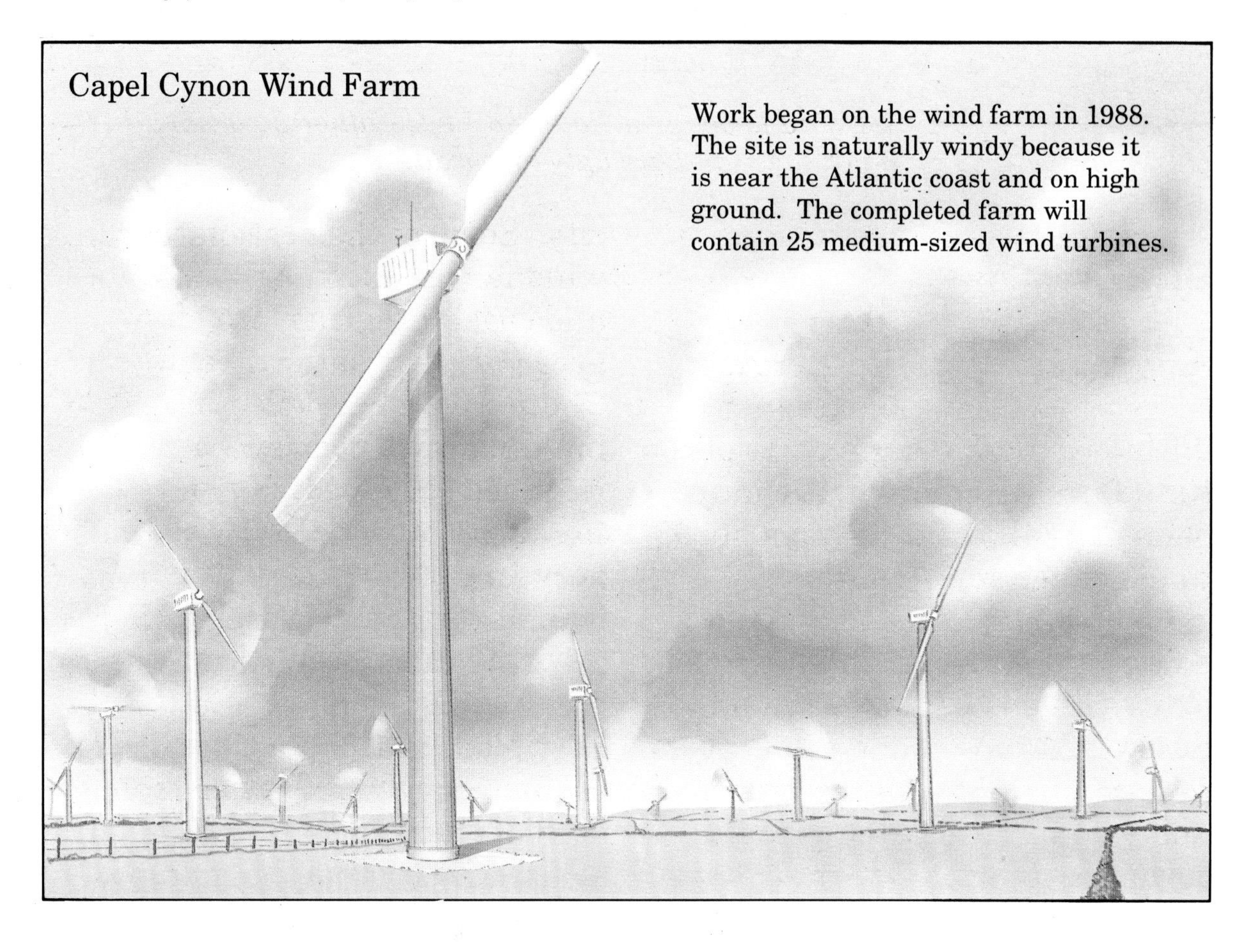

*People who live near this wind farm, near Palm Springs, California, complain that the turbines make too much noise and are ugly to look at.*

## Wind farms and the environment

As with any source of energy, there are some problems with using the wind. Some parts of the world are simply not windy enough — they will always need to use other energy sources.

Some people think that wind farms are not very pleasant to look at. The wind farm needed to replace just one coal-burning or nuclear power station would cover an area the size of a large city. Some of the best places for wind farms, such as mountains and coastal regions, are often areas of great natural beauty. These would be spoiled by building rows and rows of turbines. The whirring of hundreds of large turbines can be very noisy, and it can also interfere with local radio and television reception. Turbines may also affect birds. It is possible that the presence of turbines might stop birds from nesting, and migrating birds might also fly into the turbines.

*To prevent complaints from local people, wind turbines can be built out at sea. These turbines are on the Danish coast.*

Some of these problems could be overcome by building wind farms out at sea, where strong winds usually blow. Offshore wind farms would cost more money to build than those on land, however, and they would have to be strong enough to stand up to gales and rough seas. But because it is so much windier on the water, turbines at sea would produce far more electricity than turbines on land. So sea turbines may be worth the expense.

The world's first sea-based wind farm has already been built in Denmark. It has 16 turbines, each producing 55 kW of electricity. The wind farm stands on a jetty in the sea and also has a separate large turbine on the shore. The turbines catch the wind as it blows across the water, and they provide enough electricity for 600 homes.

In Britain, there are plans to build a single large turbine at sea, three miles (5 km) off the Norfolk coast. It should be producing electricity by the mid-1990s.

## Storing the wind's energy

As we have seen, one of the main problems with wind energy is that the wind varies all the time and sometimes does not blow at all. But there are ways to store the wind's energy until it is needed. On windy days, any excess electricity can be fed into large **batteries**. When it is calm, electricity can be taken from the batteries. On a larger scale, excess electricity can be used to pump water uphill from a low lake into a higher one. When it is not windy enough to generate

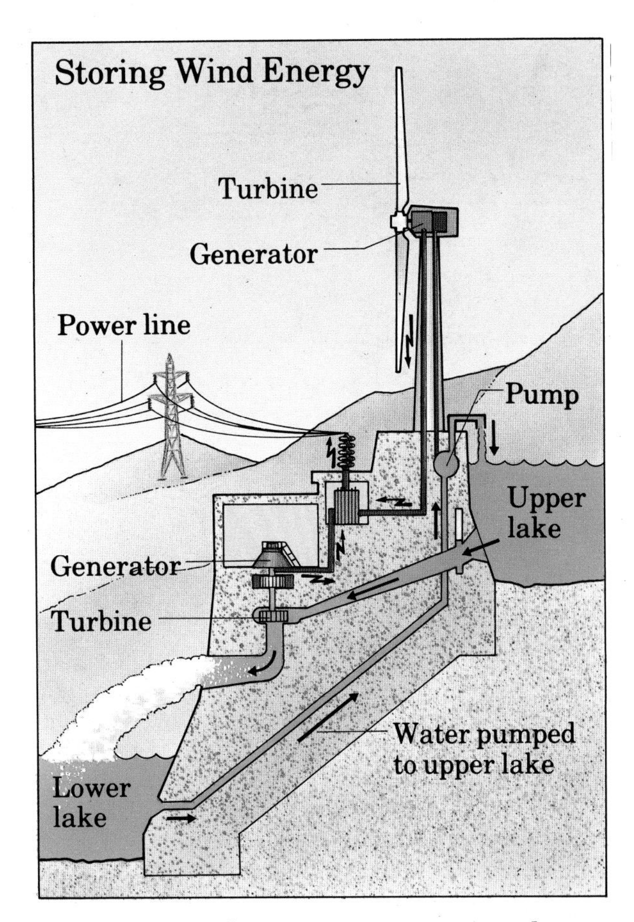


*The wind turbine pumps water to the upper lake.*

power, the water is released through water turbines, generating electricity as it runs back down to the lower lake.

If wind farms were used as a regular part of a country's main electricity system, it would not be so necessary to store wind energy. If it were windy, all wind-farm energy would be used. But if it were calm, other sources of energy would be used.

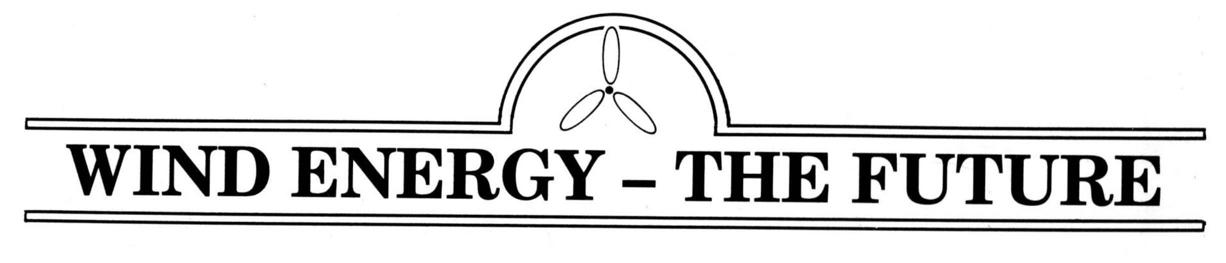

# WIND ENERGY – THE FUTURE

Wind turbines are not the only modern way to use wind energy. Modern technology is now helping wind energy to be used once again on ships. A Japanese company has recently developed a large cargo ship with tall, rectangular metal sails. The direction of the sails is automatically controlled by computers to make the best use of the wind. The ship also has an engine, but using its sails can reduce the amount of fuel that is used by 17 percent. As fossil fuels become more scarce and expensive, more and more ships may be powered by the wind.

*Sails are reappearing on modern ships. The sails on this Japanese cargo ship are controlled by electric motors and save a lot in fuel costs.*

*This device uses the power of the wind and the Sun to provide electricity for a powerful light.*

Worldwide, wind turbines now produce enough electricity for the needs of 1.25 million people. The amount of energy that we get from the wind is rising all the time. Nowadays, Europe is the center of attention for the wind-energy industry. Many countries in Europe are planning to build wind farms during the 1990s. Denmark plans to generate 10 percent of its electricity with wind turbines by the year 2000. Improvements in the way turbines are made have encouraged more people around the world to use them. India plans to use wind turbines to supply electricity to at least 5 million people by the end of this century. Other countries are likely to do the same. Wind energy will probably be one of the world's major sources of energy in the twenty-first century.

*One of the latest turbine designs — a cross-arm rotor.*

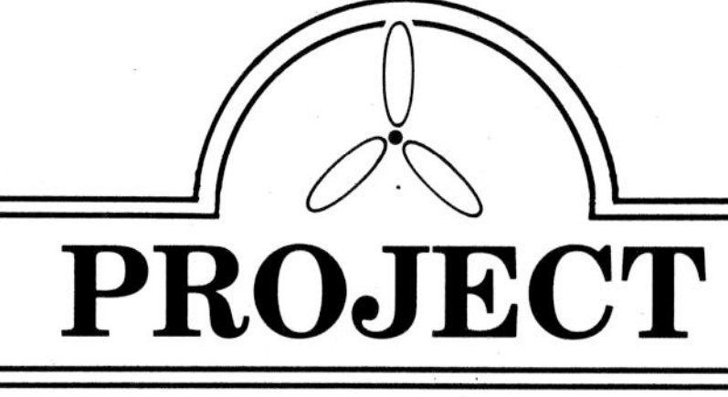

# PROJECT

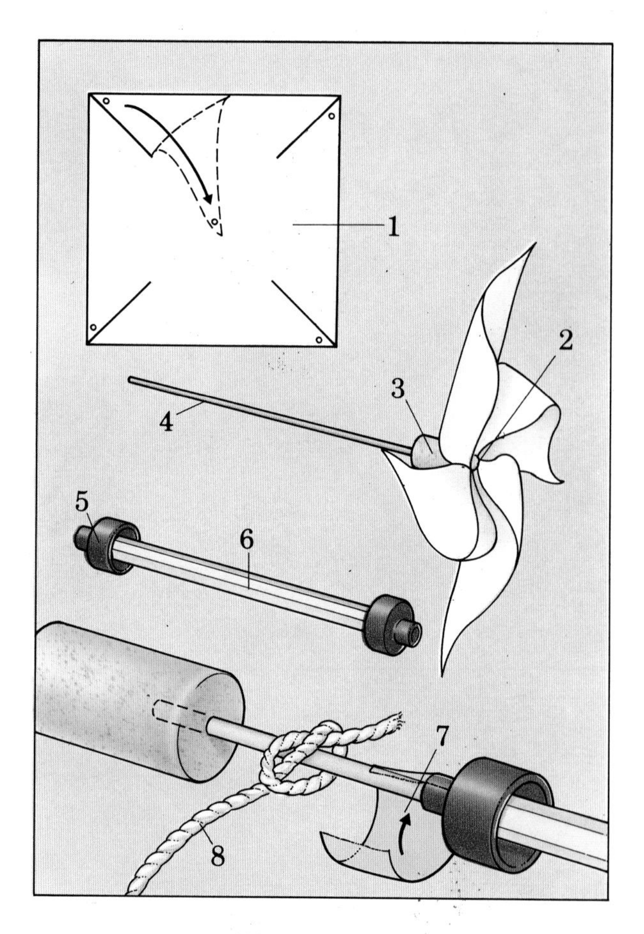


## You will need:

- A ruler
- A 12-inch (30-cm-) square piece of cardboard
- Coat-hanger wire
- A ballpoint pen casing
- Two squeeze-bottle tops
- A thumbtack
- Three corks
- String, tape, and glue

## How to make your own wind turbine:

1. Use a ruler to draw lines from corner to corner on your square piece of cardboard.

2. Make small pinholes near each corner (see diagram) and where the lines meet in the center.

3. Cut halfway along each line, starting from the corners. Bend the four corners over so that the four holes line up with the center hole. Use glue and a thumbtack to attach the center of the windmill to the end of a cork.

4. Ask an adult to cut a piece of wire about 12 inches (30 cm) long. Insert it into the end of the cork.

5. Slide one bottle top along the wire up to the cork. Slide the pen casing up to that bottle top. Slide the second bottle top up to the pen casing. Tape the second bottle top in place. Make sure that the pen casing spins freely.

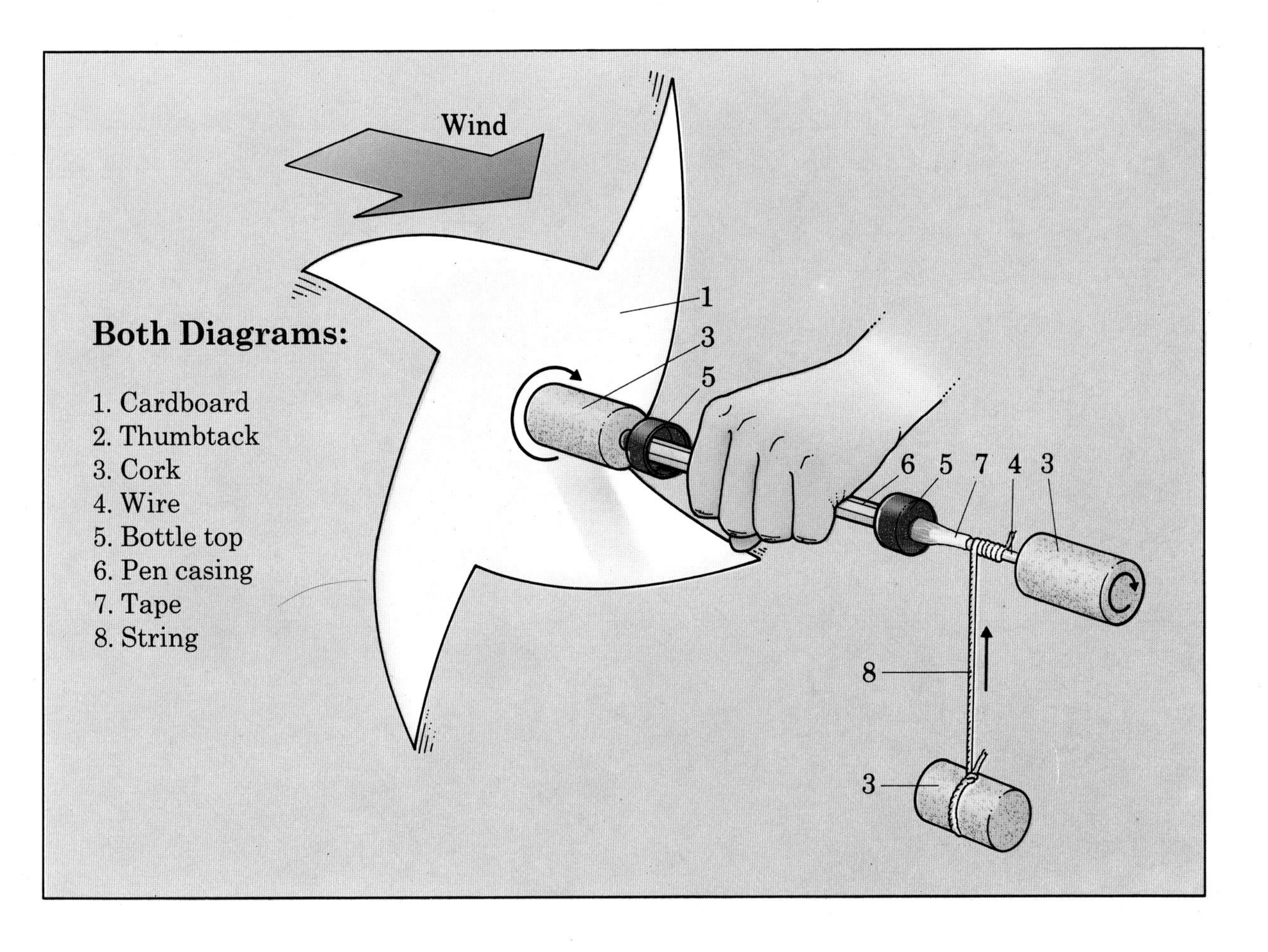


6. Tie and tape a string to the wire, about three inches (8 cm) from the free end.

7. Ask an adult to put another cork on the free end of the wire.

## Using your wind turbine:

Tie a third cork to the string. Holding the pen casing, point the turbine into a stiff breeze. It should spin, lifting the cork on the string.

If the turbine does not spin, make sure that the wire can turn freely. Try using it in a stronger wind with lighter objects tied to the string.

Your wind turbine uses wind energy to lift weights in the same way that big wind turbines use wind energy to make electricity.

# Glossary

**Acid rain:** Rain that becomes toxic from air pollutants. It kills trees and fish and, in time, can even eat away stone.

**Anemometer:** A "wind meter," used to measure how fast the wind is blowing.

**Atmosphere:** The layer of gases that surrounds a planet, moon, or star.

**Axis** (plural: **axes**): The line around which something spins.

**Battery:** A device for storing electricity.

**Carbon fiber:** A strong, lightweight material made of pure carbon.

**Clipper:** A fast sailing ship of the mid-nineteenth century.

**Cylinder:** A round tube that is often hollow and open at the ends.

**Diesel fuel:** A kind of oil used as fuel for diesel engines.

**Fiberglass:** A strong, lightweight material made of spun glass. It is used as insulation or instead of plastic.

**Fossil fuels:** Energy sources, such as coal, oil, and natural gas, formed from the remains of animals that lived millions of years ago.

**Generator:** A machine that generates, or produces, energy.

**Greenhouse effect:** The warming of the Earth's climate due to gases in the atmosphere that trap the Sun's heat.

**Grindstone:** In a mill, the stone used to grind grain into meal.

**Horizontal:** From side to side, or parallel to the horizon.

**Kilowatt (kW):** A thousand watts (see **Watt**).

**Kinetic energy:** The energy that is contained in all moving objects.

**Megawatt (MW):** A million watts (see **Watt**).

**Miller:** Someone who owns or operates a mill.

**Mountain pass:** A low-lying gap between two mountains.

**Network:** A system of connected electricity lines.

**Nuclear power:** Energy produced by splitting or combining atoms.

**Persia:** The ancient name of the country in the Middle East that is now known as Iran.

**Trade wind:** One of the Earth's major winds that blows constantly toward the equator.

**Turbine:** A device, shaped somewhat like an airplane propeller, which can spin to power an electric generator.

**Vertical:** Straight up and down, or upright.

**Watt (W):** A unit of electrical power. Most light bulbs are powered on 60 to 100 watts.

## Books to Read

*Catch the Wind. A Book of Windmills and Wind Power.* Dennis Landt (Macmillan)
*Small Energy Sources: Choices that Work.* Augusta Goldin (Harcourt Brace Jovanovich)
*A Treasury of Windmill Books.* Charles Addams and Jose Aruego (Julian Messner)
*What Makes the Wind?* Laurence Santrey (Troll)
*Wind: Making It Work for You.* Douglas R. Coonley (Laurence Erlbaum)
*Wind and Water Energy.* Sherry Payne (Raintree)
*Wind Power.* Mike Cross (Franklin Watts)

## Places to Write

These groups can help you find out more about wind energy and alternative energy in general. When you write, be sure to ask specific questions, and always include your full name, address, and age.

**In the United States:**

**Conservation and Renewable Energy Inquiry and Referral Service**
P.O. Box 8900
Silver Spring, MD 20907

**American Wind Energy Association**
777 North Capitol Street NE
Washington, DC 20002-4226

**In Canada:**

**Efficiency and Alternative Energy Technology Board**
**Department of Energy, Mines, and Resources**
580 Booth Street, 7th Floor
Ottawa, Ontario K1A 0E4

**Canadian Wind Energy Association**
44A Clarey Avenue
Ottawa, Ontario K1S 2R7

# Index

A **boldface** number means that the entry is illustrated on that page.